I0814243

WILD STEM

STEM UNDERGROUND

Megan Borgert-Spaniol

Checkerboard Library

An Imprint of Abdo Publishing
abdobooks.com

abdobooks.com

Published by Abdo Publishing, a division of ABDO, PO Box 398166, Minneapolis, Minnesota 55439.

Printed in the United States of America, North Mankato, Minnesota
102023
012024

Design: Denise Hamernik, Mighty Media, Inc.
Production: Mighty Media, Inc.
Editor: Anna Anderhagen
Cover Photographs: mihtiander/Getty Images
Interior Photographs: AZ68/iStockphoto, p. 11; benedek/iStockphoto, p. 21; Brocreative/Shutterstock Images, p. 5; fikretozk/iStockphoto, p. 7; gorodenkoff/iStockphoto, p. 19; Jane Rix/Shutterstock Images, p. 23; Mighty Media, Inc., pp. 8–9, 26–27; Paul Williams/Flickr, p. 25; PeopleImages.com - Yuri A/Shutterstock Images, p. 29; subtik/iStockphoto, p.15; 36clicks/iStockphoto, p. 17; Traci Hardin/iStockphoto, p. 13
Design Elements: Lorthois Yuliya/Shutterstock Images; supanut/Adobe Stock

Library of Congress Control Number: 2023939334

Publisher's Cataloging-in-Publication Data
Names: Borgert-Spaniol, Megan, author.
Title: STEM underground / by Megan Borgert-Spaniol
Description: Minneapolis, Minnesota : Abdo Publishing, 2024 | Series: Wild STEM | Includes online resources and index.
Identifiers: ISBN 9781098292003 (lib. bdg.) | ISBN 9781098278908 (ebook)
Subjects: LCSH: Science--Study and teaching--Juvenile literature. | Technology--Study and teaching--Juvenile literature. | Engineering--Study and teaching--Juvenile literature. | Mathematics--Study and teaching--Juvenile literature. | Rocks--Juvenile literature. | Earth sciences--Juvenile literature. | Geology--Juvenile literature.
Classification: DDC 577.0--dc23

CONTENTS

WILD UNDERGROUND

Do you love learning what happens beneath Earth's surface? Do you dream of digging for fossils or climbing through caves? If so, you might enjoy a career that takes you underground!

Paleontologists uncover buried evidence of ancient life. Speleologists explore caves and their **inhabitants**. Volcanologists study volcanoes, while seismologists record earthquakes. **Geothermal** engineers capture Earth's heat for electricity and more.

What kind of scientist would you like to be? There's only one way to find out. Put on a hard hat and grab a headlamp. It's time to explore science, technology, engineering, and math (STEM) underground!

STEM
Science
Technology
Engineering
Math

The farthest humans have drilled into Earth is 7.6 miles (12.2 km). The distance to Earth's core is nearly 240 times farther, at 1,802 miles (2,900 km)!

SEISMOLOGISTS

Have you ever felt the ground below you shake? This trembling might have been caused by passing trucks or construction equipment nearby. But it's possible you were feeling an earthquake!

Earthquakes happen when huge pieces of Earth's crust suddenly shift. This movement releases energy. The energy travels to Earth's surface through vibrations called seismic waves.

Scientists who study seismic waves are called seismologists. These researchers measure surface vibrations using **seismographs**. Seismograph data reveals to scientists where an earthquake originated and how strong it was. This research helps scientists predict when earthquakes might happen!

Most earthquakes are too weak to be felt. But some are strong enough to tear apart roads and destroy buildings.

GET WILD!

DIY SEISMOGRAPH

Create a seismograph using household materials!

WHAT YOU NEED:

- ✔ paper or plastic cup
- ✔ craft knife
- ✔ marker
- ✔ ruler
- ✔ string
- ✔ scissors
- ✔ shoebox without the lid
- ✔ rocks, beans, or other weights
- ✔ paper
- ✔ tape

WHAT YOU DO:

1. Ask an adult to cut a hole in the bottom of the cup that is just big enough for the marker to fit through. Then, ask the adult to cut two holes on opposite sides of the cup beneath the rim. Measure how far apart the holes are using a ruler. Thread a 12-inch (30 cm) piece of string through both holes.

2. Cut two holes in the top end of the shoebox, making the holes the same distance apart as the cup holes. Thread the string attached to the cup through each hole. Knot the strings together at the top of the shoebox.

3. Push the capped end of the marker down through the hole at the bottom of the cup. Fill the cup with weights.

4. Cut a sheet of paper into thirds the long way. Tape the strips together to form one long strip. Cut a slit at the bottom of each side of the box and pull the paper strip through both slits.

5. Uncap the marker and adjust it so its tip touches the paper. Have a helper slowly pull the paper strip through the shoebox while you hold the box in place. What does the marker draw on the paper?

6. Try shaking the box in different ways and observe what your **seismograph** records. If you run out of recording paper, make some more!

VOLCANOLOGISTS

Deep below Earth's surface is hot liquid rock called magma. Earth's crust traps this magma underground. In some areas, openings in Earth's crust allow magma to rise to the surface. These openings are called volcanoes. Scientists who study volcanoes are called volcanologists.

Scientists classify volcanoes by how likely they are to erupt. Active volcanoes have erupted recently and are likely to erupt again. Dormant volcanoes have not been active in many years but could erupt again. Extinct volcanoes are not expected to erupt again.

Volcanologists collect rock samples from extinct and dormant volcanoes. They study the physical and chemical makeup of the rocks. These rocks help them understand how volcanoes erupt.

In a volcano, fresh lava can be more than 2,000 degrees Fahrenheit (1,093°C)!

Volcanologists use various tools to **monitor** active volcanoes. An instrument called a spectrometer identifies the gases coming out of a volcano. A tiltmeter detects changes in the slope of a volcano. These tools alert scientists to when a volcano may erupt.

Some volcanic eruptions are calm. If the lava is flowing slowly, scientists can scoop samples of the liquid rock to study. But some eruptions are explosions of lava, ash, and rock. Scientists alert local officials if such an event is possible. These alerts give people who live near volcanoes time to evacuate the area.

Volcanologists also study how magma behaves. This research helps predict when and how a volcano will erupt. This knowledge can save lives!

There are about 1,350 active volcanoes in the world. This volcanologist inspects flowing lava from a volcano in Hawaii.

GEOTHERMAL ENGINEERS

Heat from inside Earth escapes through volcanoes. It also escapes through hot springs and **geysers**. These are places where hot water and steam rise to Earth's surface.

The water and steam come from underground **reservoirs** heated by magma. These reservoirs are sources of **geothermal** energy. Geothermal engineers are scientists who design methods that capture and use geothermal energy.

One way engineers capture geothermal energy is with heat pumps. Pipes carry water from underground reservoirs. Geothermal energy heats the water, which is then pumped to the surface. This hot water heats buildings during cool seasons.

Iceland's Strokkur Geyser is one of the most active geysers in the world. It erupts every five to ten minutes when hot water shoots up to 200 feet (61 m) into the sky!

Geothermal engineers have also created ways to turn geothermal energy into electricity. Geothermal power plants harness steam power that comes from the underground **reservoirs**. The steam energy forces **turbines** to rotate. The rotating turbines generate electricity!

Many areas of the world do not have access to underground reservoirs of hot water. In these places, engineers drill deep wells. They pump water into the wells. The water absorbs Earth's heat. Then the water is pumped up to the surface and used to create steam power.

Drilling and pumping water into the ground allows more countries to use geothermal energy. However, this practice can also create earthquakes. Geothermal engineers are always working to improve how we harness the energy of Earth's heat.

Geothermal power plants produce electricity 24 hours a day. The United States generates the most geothermal electricity in the world, powering about 2.7 million homes.

PALEONTOLOGISTS

Besides digging underground for energy resources, scientists dig to find clues about Earth's history. These clues come in the form of fossils. Fossils are preserved remains of ancient plants and animals found underground or at Earth's surface. Scientists who look for and study these fossils are called paleontologists.

Paleontologists **specialize** in certain types of fossilized organisms. Some focus their research on **vertebrates**, including birds, fish, and reptiles. Others study **invertebrates**, such as dragonflies and snails.

Discovering fossils is an important part of being a paleontologist. Paleontologists first remove any rock covering the fossil using hand tools such as hammers, **chisels**, and brushes. Next, they **excavate** the fossil along with some of the surrounding rock. They wrap

Two paleontologists use brushes to remove rock covering dinosaur bones and teeth.

the fossil in cloth, toilet paper, or plaster to keep it safe. Then, the fossil is transported to a lab.

In labs, paleontologists use more equipment. Air scribes are small power tools that break up rock like mini jackhammers. These tools are precise enough to remove rock from a fossil one grain at a time!

Imaging technologies allow scientists to study fossils in greater detail. Researchers use powerful X-rays to scan fossils. Then they use the scans to create digital 3D models of the fossils. Scientists can even create lifelike models of ancient plants and animals based on their fossils. This technology gives paleontologists a clearer picture of the history of life on Earth.

When using an air scribe, paleontologists wear masks and goggles for protection from stone dust.

SPELEOLOGISTS

Some scientists dig or drill underground to find what they are looking for. Others find places where the ground opens for them! Speleologists are scientists who study caves.

Many speleologists study how caves form. Most caves form where the ground is made of limestone or other rocks that **dissolve** in water. Over time, rainwater seeps into small cracks in the rock. As the rock dissolves, the cracks get bigger and bigger. After thousands of years, a cave is formed.

Speleologists also study life inside caves. They discover ways that cave-dwelling animals adapt to life in the dark. For example, the lack of light makes sight less useful in caves. Because of the lack of light, many cave fish and salamanders are blind.

Some caves are made of ice, not rock. This cave formed when water melted from a glacier and carved tunnels through the ice!

Scientists have also found cave animals that make their own light. New Zealand is home to glowworms that cover cave ceilings in blue light!

Speleologists navigate caves wearing helmets, gloves, kneepads, and sturdy boots. They also need light! Headlamps are primary light sources. Headlamps allow scientists to use both hands for climbing, taking photos, and recording observations.

Some caves require more **specialized** gear. For example, the temperature in Mexico's Cave of Crystals can rise above 130 degrees Fahrenheit (54°C). Speleologists who want to study the crystals must wear special cooling suits. Thanks to this gear, researchers discovered **microbes** living inside the crystals!

Mexico's Cave of Crystals holds gypsum crystals as long as school buses!

GET WILD!

SUGAR CUBE CAVE

Simulate cave formation with sugar cubes and clay! The clay represents Earth's surface. The sugar cubes represent limestone rock.

WHAT YOU NEED:

- ✔ sugar cubes
- ✔ clear container with straight sides
- ✔ clay
- ✔ toothpick
- ✔ spray bottle
- ✔ warm water

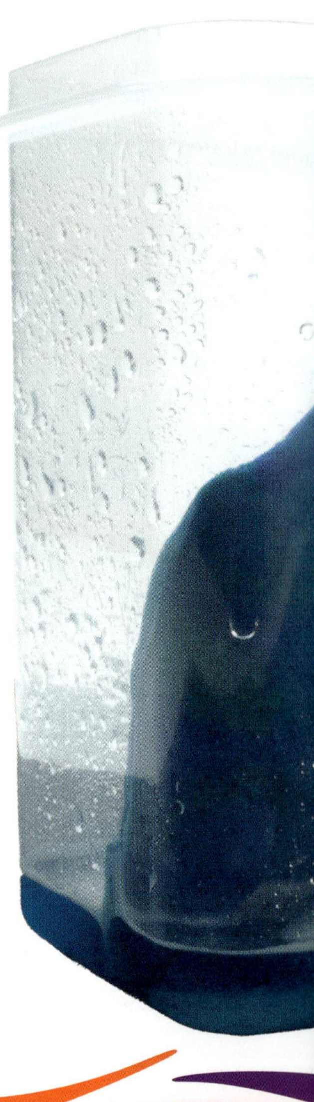

WHAT YOU DO:

1. Inside the container, construct a pyramid of sugar cubes against one side. Make sure there are no gaps between sugar cubes.

2. Flatten the clay into a disk large enough to cover the sugar cube pyramid. The clay should be about ⅛ inch (0.3 cm) thick.

3. Lay the clay over the pyramid of sugar cubes. Press the clay firmly against the side of the container so the sugar cubes are tightly sealed inside it. You should still be able to see the cubes stacked against the side of the container.

4. Use the toothpick to poke several holes through the clay.

5. Spray warm water over the clay. What happens as the water drips through the clay onto the sugar cubes?

A CAREER UNDERGROUND

Do you have a passion for STEM underground? There are many career paths to choose from. Each path leads to a wide range of opportunities.

Seismologists might collect data for governments or companies that want to dig underground for oil. Speleologists can work for mining companies or national parks that have caves. And paleontologists might do research for museums or universities. No matter where they work, these scientists share a love of what's beneath our planet's surface!

Maybe one day, your office will be underground!
What will you discover?

GLOSSARY

chisel—a metal tool with a sharpened edge at one end used to chip or cut into a solid material, such as stone.

dissolve—to cause to pass into a solution or become liquid.

excavate—to dig out and remove.

geothermal—relating to or produced by the heat from inside Earth. Geothermal energy is heat energy from Earth.

geyser—a spring that sends up occasional fountain-like jets of hot water and steam.

inhabitant—one who lives permanently in a place.

invertebrate—an animal without a backbone.

microbe—a living organism of very small size.

monitor—to watch, keep track of, or oversee.

reservoir—a natural or human-made place used for storing water.

seismograph—a scientific instrument used to measure vibrations in Earth.

simulate—to imitate.

specialize—to concentrate one's efforts in a specific area.

turbine—a device with rotating blades turned by a moving force such as water, steam, gas, or wind. A turbine turns the energy of movement into mechanical power.

vertebrate—an animal with a backbone.

ONLINE RESOURCES

To learn more about STEM underground, visit **abdobooklinks.com**. These links are routinely monitored and updated to provide the most current information available.

INDEX